AHA! ACADEMY

MAGNETIC MAGIC!

The Physics of Magnetism

Written by Kathryn Hulick

WORLD BOOK

www.worldbook.com

Co-published by agreement between Shi Tu Hui and World Book, Inc.

Shi Tu Hui
Room 1807, Block 1,
#3 West Dawang Road
Chaoyang District, Beijing 100025
P.R. China

World Book, Inc.
180 North LaSalle Street
Suite 900
Chicago, Illinois 60601
USA

Library of Congress Control Number: 2024947134

Aha! Academy: Physics
ISBN: 978-0-7166-7144-2 (set, hard cover)

Magnetic Magic! The Physics of Motion
ISBN: 978-0-7166-7151-0 (hard cover)
ISBN: 978-0-7166-7171-8 (e-book)
ISBN: 978-0-7166-7161-9 (soft cover)

Printed in India by Replika Press PVT LTD, Haryana, India
1st printing January 2025

Staff

Editorial

Vice President
Tom Evans

Editorial Project Coordinator
Kaile Kilner

Senior Curriculum Designer
Caroline Davidson

Proofreader
Nathalie Strassheim

Graphics and Design

Senior Visual
Communications Designer
Melanie Bender

Designer
Shannon Hagman

Digital Asset Specialist
Rosalia Bledsoe

Written by Kathryn Hulick
Advised by Farrukh J. Fattoyev

Developed with World Book by
Red Line Editorial

Acknowledgments

The publishers gratefully acknowledge the following sources for photography. All illustrations were prepared by WORLD BOOK unless otherwise noted.

Cover: NASA/JPL-Caltech; Denis Belitsky, Shutterstock; cyo bo/Shutterstock; Daxiao Productions/Shutterstock; Nuttapong Photographer/Shutterstock; revers/Shutterstock

blickwinkel/Alamy Images 24; Dorling Kindersley Ltd/Alamy Images 45; CSIRO (licensed under CC BY 3.0) 25; ESA/Planck Collaboration 26; Maryland GovPics (licensed under CC BY 2.0) 39; NASA 27, 28, 29, 31, 33; NASA, ESA, CSA, STScI 26; NASA/JPL-Caltech 33; NASA/SDO/AIA/LMSAL 29; Oak Ridge National Laboratory (licensed under CC BY 2.0) 43; Public Domain 20, 21, 23, 39; Johannes Reimer (licensed under CC BY 4.0) 43; Shutterstock 3, 4, 5, 6, 7, 8, 9, 10, 11, 12, 13, 14, 15, 16, 17, 18, 19, 20, 21, 22, 23, 24, 25, 26, 27, 28, 29, 30, 31, 32, 33, 34, 35, 36, 37, 38, 39, 40, 41, 42, 43, 44, 45, 46, 47, 48; Victoria C (licensed under CC BY 4.0) 23

There is a glossary of terms on page 48. Terms defined in the glossary are in type that looks like *this* on their first appearance on any spread (two facing pages).

Contents

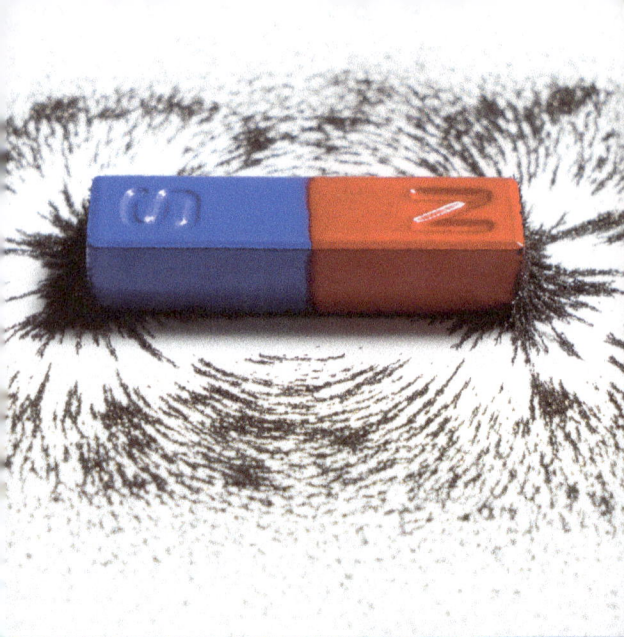

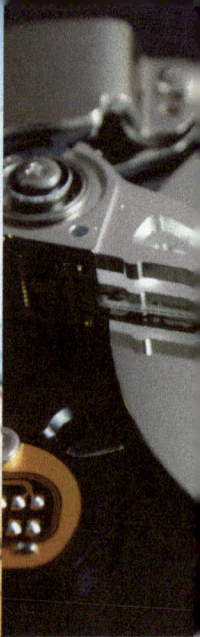

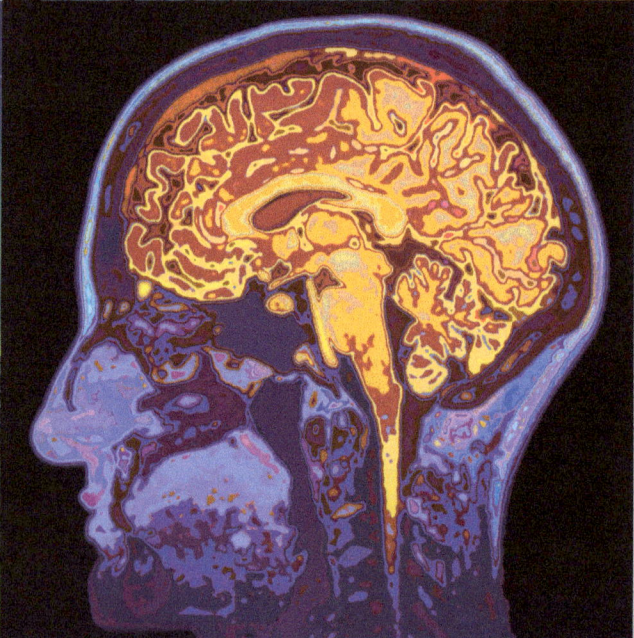

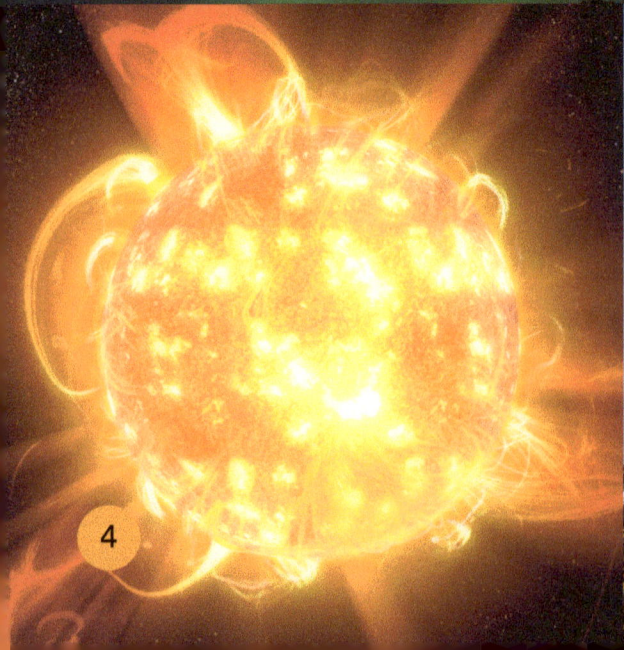

Introduction

What can make a train float? What can move a spoon without touching it? The answer isn't magic—it's magnets!

Magnets work thanks to a *force* called magnetism. This force is both completely ordinary—it's how you stick a photo to a refrigerator—and completely extraordinary.

Magnetism holds a computer's memory, helps a sea turtle find its way home, shapes the surface of the sun, and so much more. Let's learn the secrets of this amazing force!

Am I there yet?

What do this train and this turtle have in common? They both rely on magnetism!

1

WHAT IS MAGNETISM?

Every magnet has a north pole and a south pole. Matching poles repel each other—kind of like they each have a stinky smell that keeps the other away!

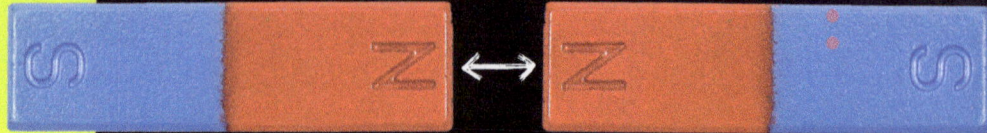

S N ↔ N S

Meanwhile, opposite poles attract each other—like they're both wearing a sweet-smelling perfume that draws them to each other.

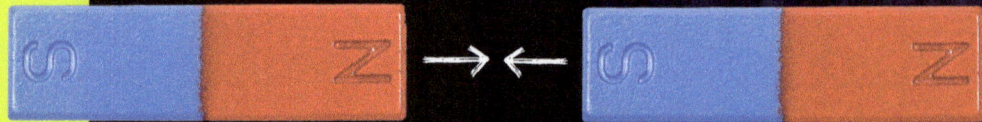

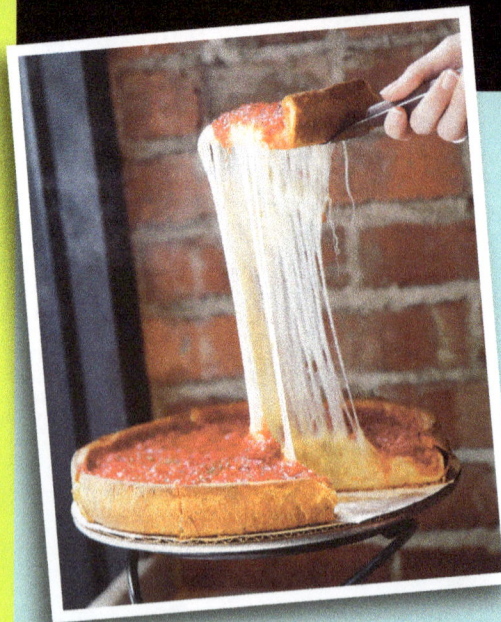

S N → ← S N

Magnetism is an invisible _force_ that can do amazing things. It's the reason a magnetic hoverboard floats. It also makes such gadgets as headphones and computers work!

A magnetic force is stronger when magnetic objects are closer together. Think about odors again. The smell of something delicious, like pizza, can really draw you in—but not if you're too far away to smell it. The more distance between magnetic objects (and between you and your pizza), the weaker the force.

The Lexus hoverboard was an invention that looked kind of like the one pictured here. It was like a skateboard, only it floated just above the ground! It only worked at one special park in Spain that had powerful magnets built into the ground. Pro skateboarders helped test it out.

Check out these magnetic moves!

Magnetism and electricity are like two sides of the same coin. In physics, they are two parts of one force called *electromagnetism*. You'll see how magnetism and electricity work together later in this book. But first, let's explore how a magnet works!

Magnetize!

Imagine electrons as tiny twirling dancers. They have a property called spin, which makes magnetism. You might say an electron is the world's tiniest magnet! The direction of an electron's spin determines where its north and south poles are. Think of a dancer spinning either to the left or to the right.

Wheee!

In most atoms, electrons come in pairs with opposite spins. They both make some magnetism, but it all cancels out. Some atoms, though, have something called an unpaired electron. Imagine a solo star dancer really getting going as they spin around and around. It gives the whole atom a tiny bit of magnetism.

Where does magnetism come from?

It all starts with *electrons*, the charged particles that swirl around the outside of an *atom*.

In the real world, atoms come in groups.

When lots of atoms with those solo star electrons get together, some groups of electrons will spin in sync, forming *magnetic domains*. These are like mini magnetic dance troupes.

Huh? Which way are we spinning?

Most of the time, these domains (dance troupes) can't agree to spin in the same direction.

When they spin in opposite directions, they cancel each other out. The result is very weak magnetism, or no magnetism at all.

Can something make the troupes dance together?

Yes! Electricity or another magnet can align the magnetic domains. Now, their magnetism all adds together instead of canceling out. The material they are all a part of now has an invisible *force* field strong enough to notice. It will stick to magnetic objects!

That's so metal

Nickel

Cobalt

Iron

Nickel, cobalt, iron, and some other metals have a special property called ferromagnetism.

The *atoms* in these metals are organized into a regular structure. Also, the atoms have unpaired *electrons* (those solo star dancers). Once the dance troupes in these metals get aligned and spinning in the same direction, they tend to stay that way. This forms a permanent magnet, like the kind you stick to a refrigerator.

Ouch!

DID YOU KNOW?

The strongest known permanent magnets contain the metals neodymium, iron, and boron. Online videos show these magnets attracting each other with so much *force* that they crush juice cartons, eggs, fruit, and more!

A magnet sticks to a metal refrigerator door.

But it won't stick to skin, plastic, paper, or many other materials. What's so special about the metal door?

What about the refrigerator?

Most fridge doors are made of steel, which contains iron. When a magnet comes close, the steel door's dance troupes sync up. But the other metals mixed in with the iron are more disorganized. If there's no magnet nearby, the troupes go back to their disordered ways. So the door is a temporary magnet—that is, it is only magnetic when a permanent magnet is nearby.

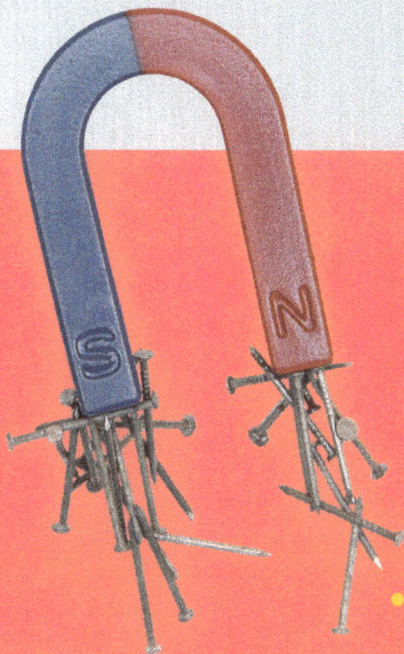

Paper clips, nails, keys, silverware, and many other common metal objects are also temporary magnets. Metal spoons in a drawer don't usually stick to each other. But they will stick to a magnet. And a magnet can make them temporarily stick to each other.

Invisible force fields

What's doing the pushing and pulling?

It's a magnetic field! This field is invisible to people. But magnetic materials and electrical charges react to it. Like trains on a track, they must follow its curved lines.

Remember, every magnet has two poles, and opposite poles attract. The lines of a magnetic field start at the north pole and follow a curved path to the south pole. Within the magnet, the lines move from the south pole to the north pole.

Magnetism seems magical. In fact, magicians use magnets in some magic tricks. If a ball seems to float or move by itself, there's a good chance magnets are involved. A magnet can reach through empty space to push or pull on nearby objects!

Magic? Or magnets?

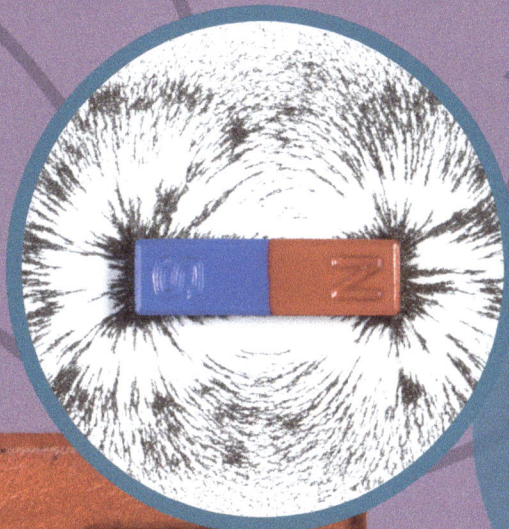

If you sprinkle bits of iron over a bar magnet, you can see the lines of the magnetic field!

Some magnetic fields are stronger than others. A weak refrigerator magnet holds up a thin sheet of paper but not a thick, folded greeting card. Its magnetic field has a strength of around 10 to 50 gauss. That's a unit of measurement named after the mathematician Carl Friedrich Gauss. Ten thousand gauss equals one tesla. This unit is named after the inventor Nikola Tesla.

Electromagnets

Electromagnets have a few simple ingredients.

Electricity
A battery or other source provides power.

Coiled wire
The wire coil, usually made of copper, conducts electricity.

Metal core
The wire wraps around the metal.

In devices called electromagnets, an electric current creates a magnetic field.

Electromagnets are inside all kinds of everyday gadgets, including headphones, computers, vacuum cleaners, and more. But how do they work?

Rock on!

Why does this work? As electricity flows through a straight wire, *electrons* inside all move in sync. This creates a magnetic field around the wire, but it's very weak. Wrapping a wire into a tight spiral packs all those moving electrons close together. This boosts the strength of their magnetism. Adding a metal core in the middle provides even more magnetism. That's because the unpaired electrons in the metal join the dance party.

TECH TIME

Electromagnets can be superstrong. In fact, some can lift entire cars! Junkyards use these powerful electromagnets to sort through large pieces of scrap metal. Electromagnets lose their magnetism when their electricity is turned off. That's how junkyard electromagnets drop off what they've picked up!

2

MAGNETISM IN NATURE

Many amazing natural events occur thanks to magnetism. This *force* can make colorful lights dance in the night sky, guide animals on long migrations, and more. In fact, planet Earth itself is one huge magnet!

Deep inside Earth, a core of liquid iron swirls around. The churning core creates currents of electricity. These currents produce a huge magnetic field that surrounds the entire planet! This is called a magnetosphere.

Pardon my wind!

Solar wind

Earth's magnetosphere isn't perfectly round. A stream of charged particles called solar wind zips from the sun and squishes one side of the magnetosphere while stretching out the other side.

The geographic north and south poles mark out an invisible axis around which the planet rotates. The magnetic north and south poles are on the opposite sides of their geographic counterparts. They also aren't lined up exactly with the geographic poles.

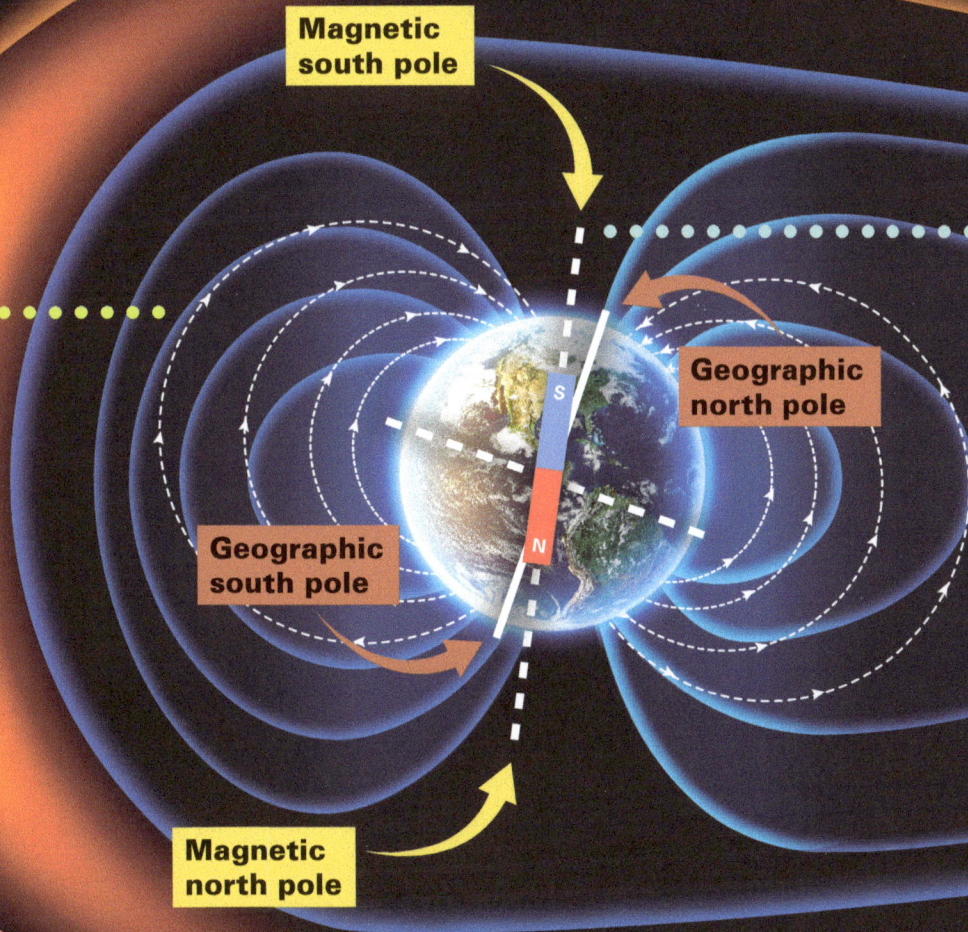

Magnetic south pole

Geographic north pole

Geographic south pole

S

N

Magnetic north pole

Magnetism helps protect the planet.
If Earth had no magnetosphere, solar wind would strip away the atmosphere and life wouldn't exist!

Dancing sky lights

The northern lights are called the aurora borealis. They often appear over northern Canada, Alaska, Scandinavia, and Russia.

There are southern lights, too! These are called the aurora australis and can be seen over New Zealand, Tasmania, and Antarctica.

Why do auroras appear? The story starts with solar wind. As these charged particles approach Earth, they follow the lines of the planet's magnetic field. The magnetic field sends the particles toward Earth at the North and South poles.

You look up and see a curtain of colored light dancing in the night sky.
This stunning natural light show is called an aurora. And it happens because of magnetism!

When they reach Earth's atmosphere, the particles crash into *atoms* of gas.
This happens in the sky above each pole, about 60 to 155 miles (100 to 250 kilometers) above the ground. The energy from each crash lights up the gas. Oxygen produces a greenish-yellow or red light. And nitrogen produces blue and purple!

Wow!

Made by lightning

Early cultures in *Mesoamerica* may have been the first to use magnetism. The Olmec people likely used lodestone as a tool more than 3,000 years ago. Starting around 500 B.C., the Monte Alto people of the same region carved rocks into human figures. These had naturally magnetic areas in their faces or belly buttons.

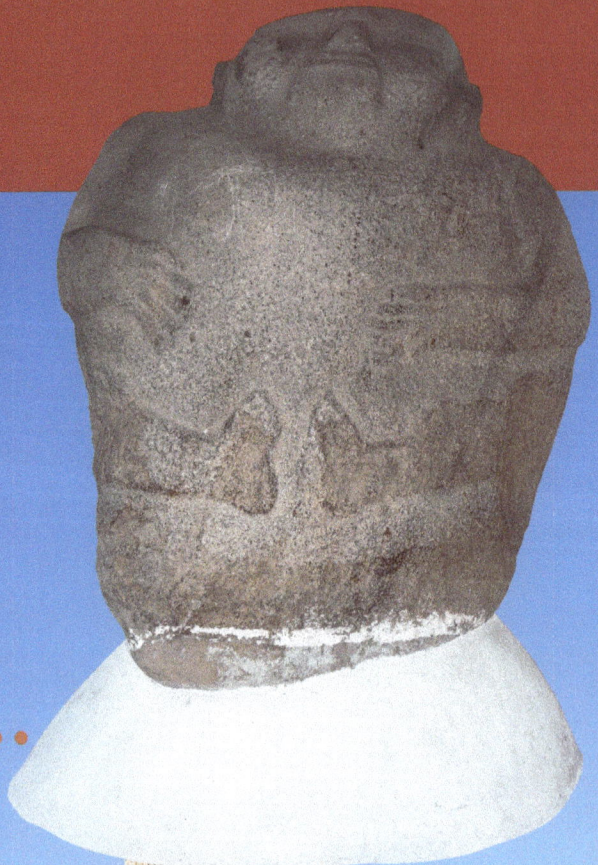

Magnets are alive.

Magnetism also fascinated the ancient Greeks. Thales of Miletus was a Greek philosopher who sought to explain natural events. He thought lodestone had a soul because it could move iron.

Ancient people discovered magnetism in rocks. These natural permanent magnets are called lodestones. They are a special form of magnetite, a hard, black, shiny mineral.

Today, scientists know that lightning creates lodestones. Lightning strikes with around 300 million *volts* of electricity. This generates a brief but powerful magnetic field that can strengthen magnetite's weak magnetism. The result is a lodestone!

Some say magnetism is named after the **Greek shepherd Magnes**. Legend has it that he was herding sheep when iron nails in his sandals got stuck to a rock. The more likely source of the word is the ancient city of Magnesia, located in what is now Turkey. Magnesia was a source of lodestone for the Greeks.

21

Due **north**

The needle on a compass always points north. This happens because the needle is a magnet. When any magnet is lightweight enough and free to move, it will line up with Earth's magnetic field. Remember that Earth's magnetic south pole is located near the geographic north pole. So the north pole of the compass needle is drawn in this direction.

The ancient Chinese made the first known compasses around 200 B.C. Typically, they crafted a spoon out of lodestone with a handle that pointed south.

You're hiking in a forest and accidentally stray from the path. You stop and look around. Which direction are you facing? Thankfully, you have a compass to show you the way!

By the A.D. 1000's, the Chinese were navigating with compasses. They had learned to use lodestone to magnetize a needle. Then they would float the needle on water.

Matthew Henson, Robert Peary, and four Inuit hunters, Ooqueah, Ootah, Egingwah, and Seegloo, were likely the first people to reach the geographic north pole. A compass helped guide their way on the famous 1909 expedition.

We made it!

European navigators later used compasses to explore the world. By this time, most compasses had needles that rotated around pins.

Animal senses

After they hatch, loggerhead sea turtles make their way across hundreds or thousands of miles of open ocean. When the turtles are ready to nest, they use magnetoreception to return to the same area where they were born.

Home sweet home!

Mole rats live in dark burrows. The Ansell's mole rat is nearly blind, but its eyes sense magnetic fields. Studies suggest the mole rats use this sense to determine where to build their nests.

Some animals have what seems like a superpower. It's called *magnetoreception*, and it's like an internal compass! Many types of fish, bats, birds, insects, and even dogs have this unusual sense.

Let's walk north!

Bogong moths in Australia migrate to get away from intense summer heat. They fly more than 600 miles (965 km) to a few small caves in the mountains, using magnetoreception to find their way.

Many songbirds fly across entire continents to the exact same breeding site each year. Scientists believe these birds take cues from Earth's magnetic field to know where and when to stop migrating.

3

MAGNETISM IN SPACE

Remember, the magnetic field that surrounds Earth is called the magnetosphere. Magnetospheres also surround many planets, most stars, and even entire galaxies. Astronomers study magnetic fields to learn about the universe!

Galaxies like the Milky Way have very weak magnetospheres— usually even weaker than the magnetic field of a refrigerator magnet! But they are still strong enough to move charged gas particles around.

The lines in this image show the orientation of the Milky Way's magnetic field.

Magnetic fields influence how stars form out of vast clouds of charged gas. *Gravity* pulls star-forming material together. And magnetism channels this material in different directions.

In the search for life on other planets, a magnetosphere is one important feature to look for. That's because it shields a planet and its atmosphere from many types of harmful *radiation*.

DID YOU KNOW?

Jupiter's moon Ganymede is the only moon known to have its own magnetosphere.

Jupiter

I'm special!

Ganymede

Swirling spotted sun

The sun is made of a soup of superhot charged particles called plasma. Moving plasma creates magnetic fields.

A prominence is a feature of plasma suspended by tangled magnetic field lines. It usually looks like an arch rising far above the surface of the sun.

Sunspots are dark spots on the sun's surface. These are areas where the sun's magnetic field is strongest. The strong magnetic field makes these spots cooler and darker than their surroundings.

A solar flare is an explosion of energy that takes the form of a bright flash of light. It happens when magnetic field lines snap or split.

The sun's magnetosphere swirls and twists. Over the course of 11 years, it flips entirely, so that the north and south poles change places! Throughout this solar cycle, the sun experiences periods of high and low activity.

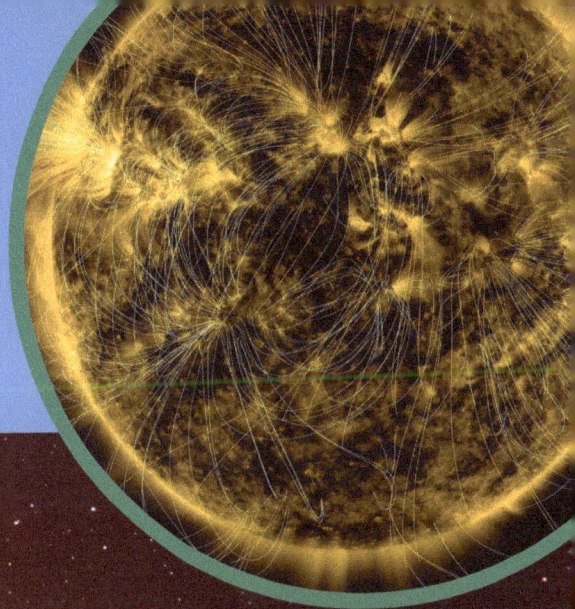

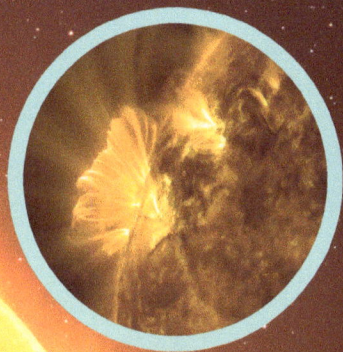

Coronal loops

tend to extend upward from sunspots. Their arching shapes trace the lines of the sun's magnetic field.

A coronal mass ejection is a huge eruption of plasma that often occurs when magnetic field lines twist until they snap like a stretched rubber band. This results in a sudden release of energy. Plasma erupts outward into space with the power of 20 million nuclear bombs.

TECH TIME

Sometimes, coronal mass ejections reach Earth's magnetosphere. This can cause electronic disruptions in satellites orbiting Earth. It can also overload power lines on Earth, leading to *blackouts.*

The solar system

Venus and **Mars** are the only two planets without a magnetic shield, though Mars likely had one billions of years ago.

Mercury has the weakest magnetic field in the solar system.

Earth's magnetic poles move slowly over time. They even swap places over the course of thousands to millions of years!

Jupiter has the strongest, most vast magnetic field of all the planets in the solar system. If it were visible, it would appear two to three times the size of the moon to stargazers on Earth!

Earth isn't the only planet with a magnetosphere. The solar system is filled with magnetic fields!

Neptune also has a crooked magnetic field. On both Uranus and Neptune, the magnetic field seems to come not from the core of the planet but from layers of ice that can conduct electricity!

Uranus rotates sideways and has a tilted, off-center magnetic field.

On **Saturn's** famous rings, shadowy streaks known as spokes sometimes appear. Astronomers believe the planet's magnetic field may cause these features.

CAREER CORNER

Planetary scientists study the planets and moons of our solar system as well as *exoplanets* that orbit distant stars. They study what distant worlds might be made of, how they form, and more. Magnetospheres are an important part of this research.

A star's second life

Neutron stars are bizarre. They are about as wide across as New York City. But the *mass* of 500,000 Earths is crammed into that space! Electrical currents flowing through neutron stars generate magnetic fields that are more than a trillion times stronger than Earth's magnetic field.

How bizarre!

When a massive star reaches the end of its life, it goes out with an explosion called a supernova. The most massive stars then collapse into black holes. Less massive stars (though still far more massive than the sun!) collapse into neutron stars.

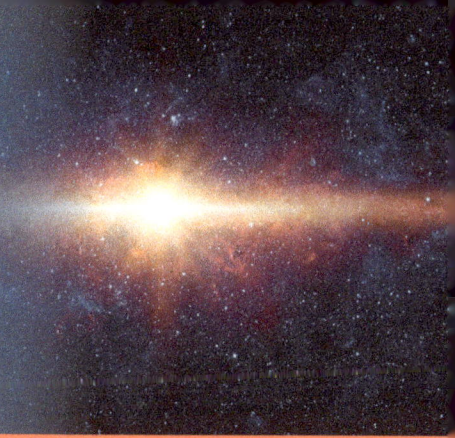

A pulsar is a neutron star that rotates rapidly. Its strong magnetic field produces twin beams of light that blast outward from each magnetic pole.

A magnetar is a special type of pulsar with an ultrastrong magnetic field. If a magnetar were to replace our moon, its magnetism could wipe out the memory of most computers on Earth!

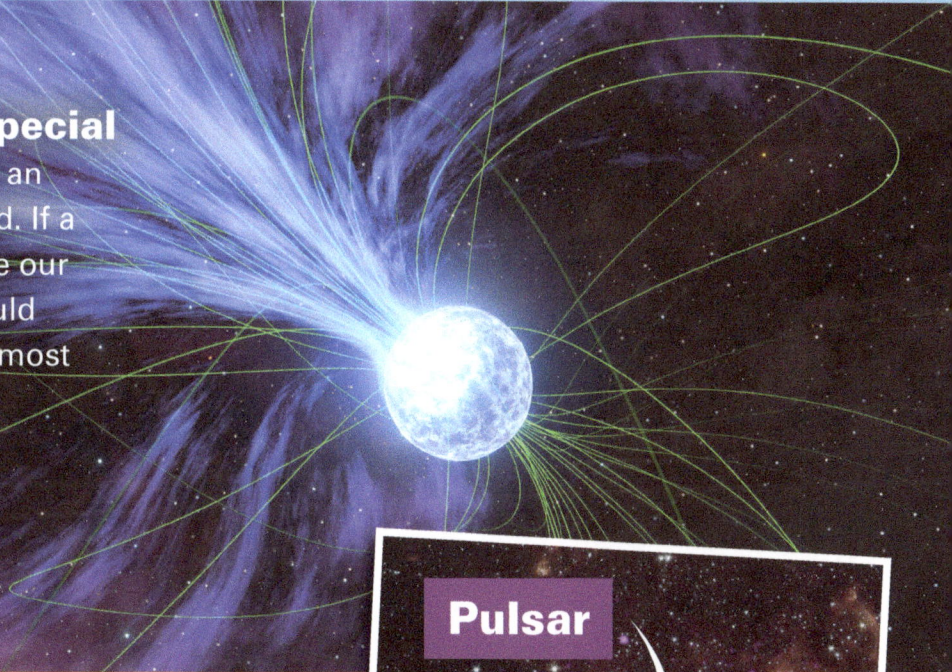

CURIOUS CONNECTIONS

ASTRONOMY A pulsar is kind of like a lighthouse. When viewed from Earth, its light beams seem to flash on and off. Pulsars flash so regularly that astronomers use them like clocks to take measurements.

Pulsar

4

MAGNETIC TECH

Most people have lots of magnets at home—and not just on the refrigerator.

Magnets are hidden inside speakers, electric toothbrushes, credit cards, and other items. They play many roles in different types of technology.

Anybody home?

Magnets also turn electricity into motion in doorbells, speakers, headphones, and other noisy gadgets. This motion vibrates a device called a diaphragm, which produces sound.

34

Remember how an electrical current flowing through a coil of wire produces a magnetic field? In the motor of an electric car, this magnetic field interacts with nearby permanent magnets. This creates a *force* that generates motion.

Older credit cards were made with a stripe of magnetic tape. The tape used the direction of magnetic fields as a code to store information. New credit cards use small chips instead because they are more secure. But many hotel door keys still use magnetic tape!

Thanks to *electromagnetism*, you can recharge a smartphone without plugging it in! Special charging pads use an electromagnetic field to transfer energy to the device.

DID YOU KNOW?

MRI stands for magnetic resonance imaging. An MRI machine uses powerful magnetic fields and radio waves to form a picture of the inside of a person's body.

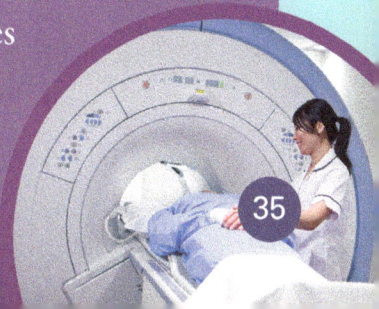

35

From motion to electricity

Geothermal power plants use heat from deep inside Earth to boil water and make steam that turns a turbine.

In hydropower plants, rushing water turns turbine blades.

Fossil fuel power plants burn coal, oil, or gas to heat water to make steam. The steam energy then spins the turbines.

Magnets help provide the world with electricity. In most power plants, an energy source spins a turbine, which is like a large fan. Then an electromagnetic generator turns this motion into electricity!

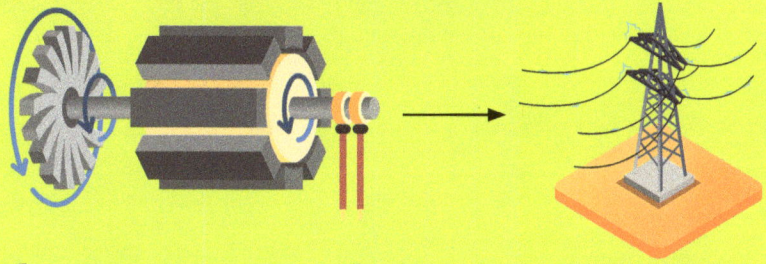

In most power plants, a spinning turbine rotates an electromagnetic shaft past a series of wire coils. This creates a moving magnetic field that makes *electrons* flow in the wire, generating an electrical current.

Some wind turbines contain large permanent magnets. As the turbine blades spin, these magnets move past coiled wire to create electricity.

DID YOU KNOW?

A wind turbine looks simple on the outside. But take a look inside one, and you'll see it requires thousands of components— up to 8,000, in fact!

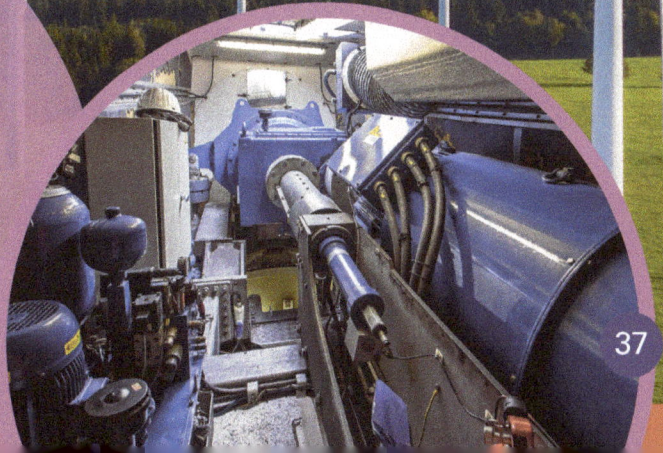

Floating trains

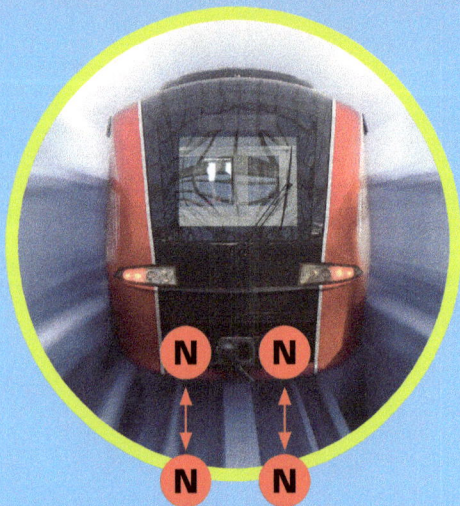

Magnets lift a maglev train so it floats a few centimeters above the track.

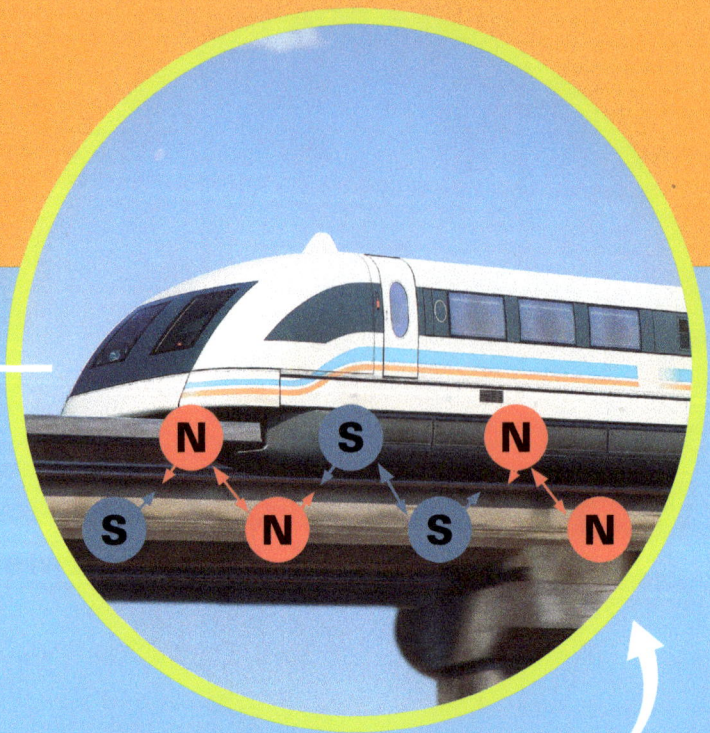

Magnets in the track of a maglev train are constantly changing their *polarity* from north to south. This creates magnetic fields in front of the train that pull it while fields behind the train push it. These *forces* propel the train forward.

The world's six maglev train systems are all located in China, South Korea, and Japan. One in China carries people from downtown Shanghai to the city's airport, a distance of about 19 miles (30 km), in under eight minutes.

Humans have yet to take regular trips in flying cars. But some can travel on floating trains! This is possible thanks to magnetic *levitation*, or maglev for short.

A maglev train in Japan set a new speed record during a 2015 test run. It reached 375 miles (603 km) per hour!

Buckle up!

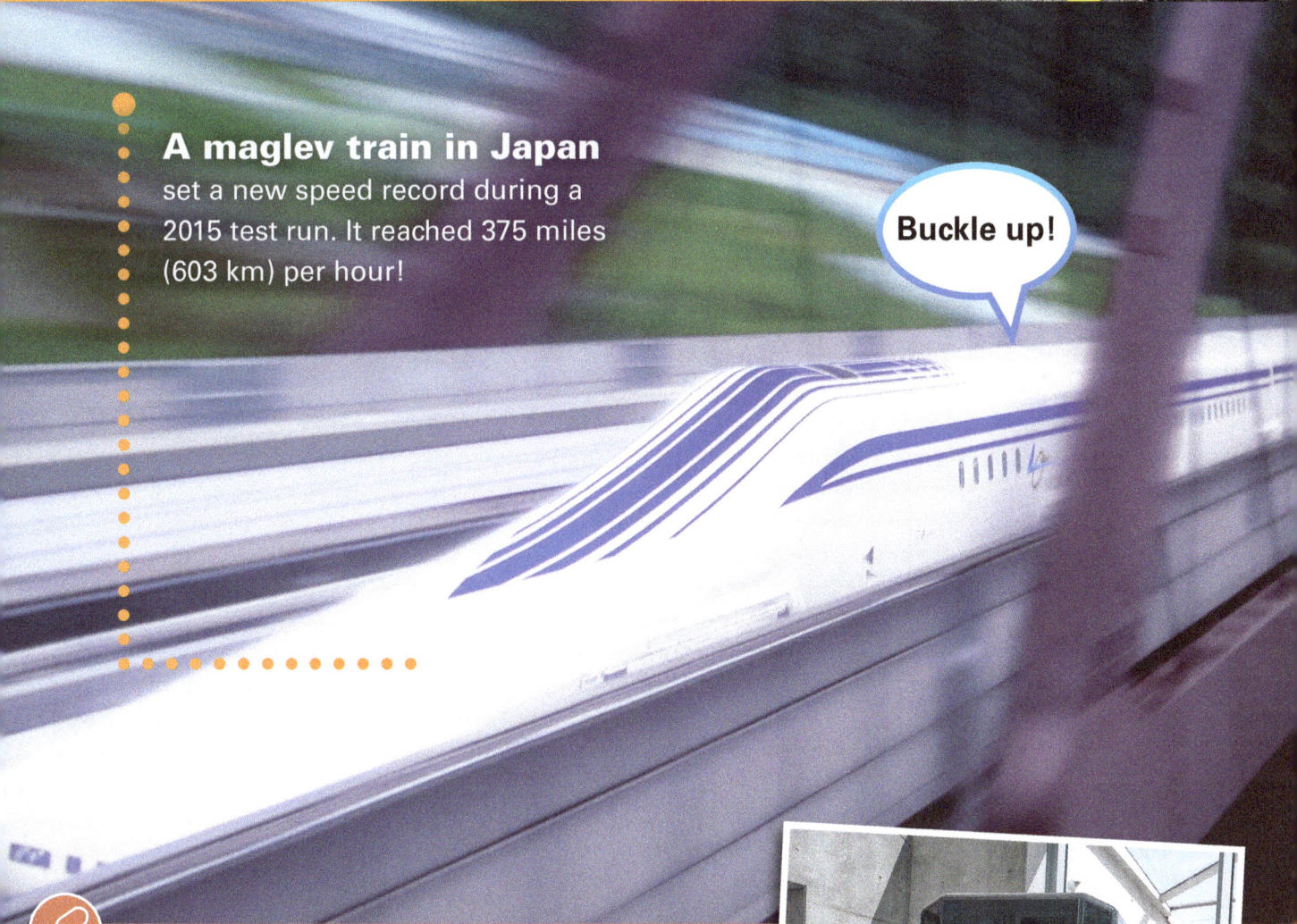

CURIOUS CONNECTIONS

ENGINEERING
Engineers are developing elevators that use maglev technology. These elevators move up and down with the help of magnets instead of cables. They can also travel sideways!

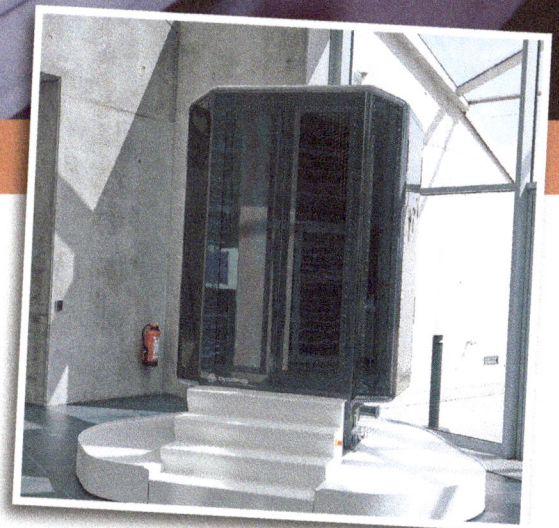

Computer memory

To save data, a computer writes electrical signals into magnetic material. To retrieve saved data, the hard drive reads magnetic information and turns it back into electrical signals.

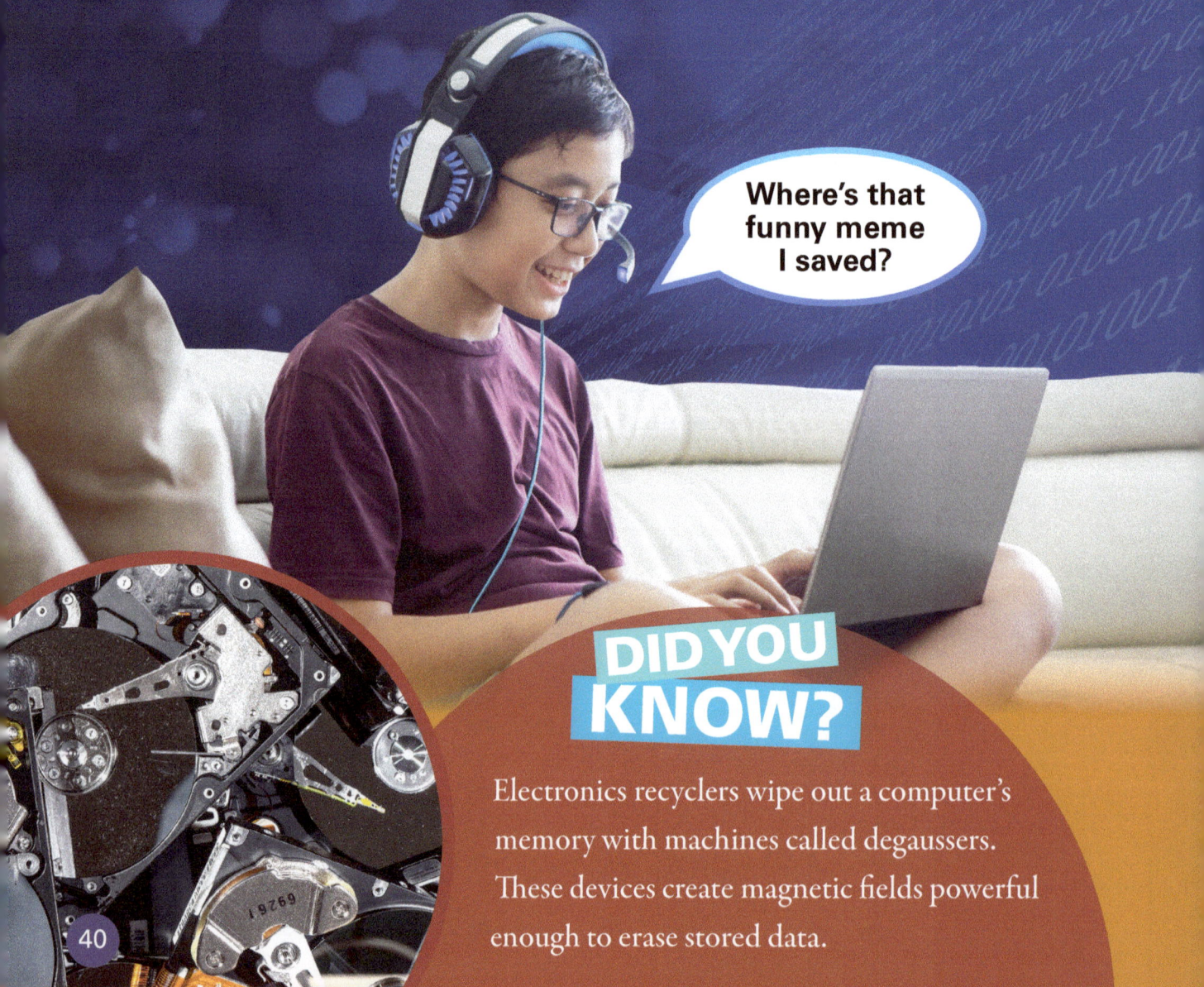

Where's that funny meme I saved?

DID YOU KNOW?

Electronics recyclers wipe out a computer's memory with machines called degaussers. These devices create magnetic fields powerful enough to erase stored data.

Computers can save everything from silly cat videos to video game progress. These data take the form of electrical signals represented by ones and zeros. A computer's hard drive stores data using magnetism!

The disk is where data are stored. It is made of a nonmagnetic material, typically glass or aluminum. Tracks of magnetic material cover the top of the disk. These tracks are so narrow that 300,000 of them can be packed into 1 inch (2.5 centimeters) of space!

The head is the part that writes electrical signals into the disk to store data. To retrieve data, the head converts magnetic information on the disk back into electrical signals.

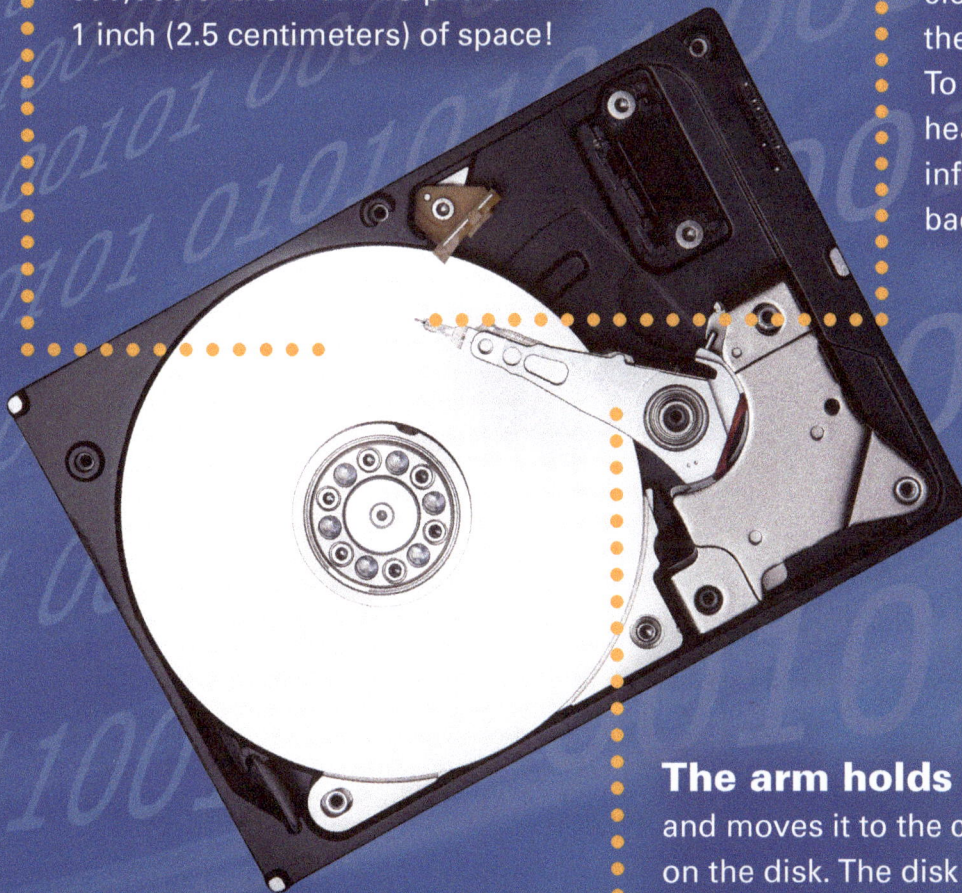

The arm holds the head and moves it to the correct location on the disk. The disk spins as the head rapidly reads or writes.

Record-breaking magnets

Superconductors are materials that can conduct electricity without any *resistance*. Coils of superconducting wire can carry so much electricity that they produce very high magnetic fields. However, known superconductors can do this only at frigid temperatures—typically around −321 °F (−196 °C) or colder!

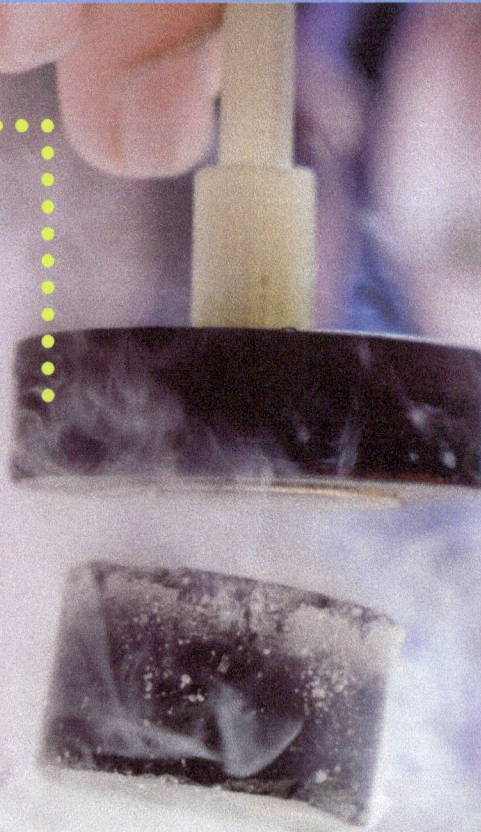

The world's largest scientific instrument is the Large Hadron Collider (LHC) in Switzerland. Tiny particles zip around this 17-mile (27-km) tunnel lined with thousands of superconducting magnets. The magnets make the particles smash together. These collisions have led to new discoveries in physics. The portion of the LHC pictured here creates a magnetic field about 100,000 times stronger than Earth's!

Extremely powerful magnetic fields, called high fields, are tricky to produce and control. But they are very useful for many types of scientific research. Superstrong magnets are also making new types of technology possible!

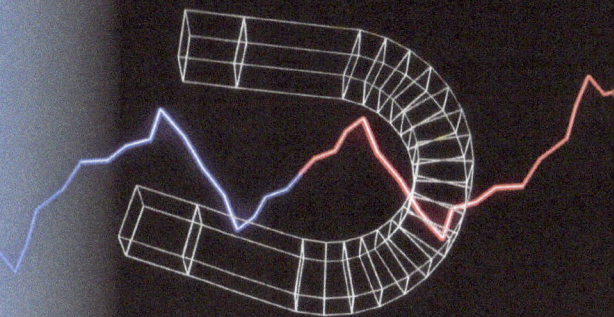

The biggest system of magnets ever made will form the heart of the *fusion* reactor ITER in France. In a fusion reactor such as ITER, heat and pressure cause *atoms* to merge. Magnetic fields form a tight cage to help keep this process going. Fusion releases lots of energy. Researchers hope to eventually make a fusion reactor work as a power plant!

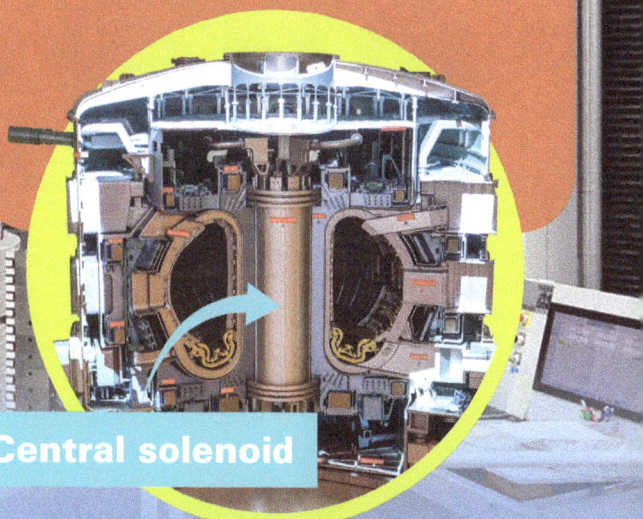

Central solenoid

Here, engineers work on the central solenoid shown in the ITER model above. This tower of superconducting magnets will reach five stories tall and weigh more than 2 million pounds (907,000 kilograms)!

CAUTION CAUTION CAUTION CAUTION

Make your own compass

You will need:

- A metal sewing needle
- A strong magnet, such as a neodymium disc magnet
- Scissors
- A piece of cork or foam (make sure it floats!)
- A small nonmetal bowl filled halfway with water

Give it a try

1. Rub the magnet down the length of the needle several times, always in the same direction. This aligns the **magnetic domains** (remember the dance troupes?) in the needle, giving the needle a magnetic field.
2. The needle should now be magnetized. Check if it sticks to the scissors or another metal object. If it does not, repeat Step 1.
3. Have an adult use the scissors to cut a piece of cork or foam. It should be about ¼ inch (0.6 cm) thick and about the size of a small coin.
4. Place the bowl of water on a flat surface. Check that the piece of cork or foam floats on the water. Then take it out.
5. Have an adult carefully stick the needle lengthwise through the cork or foam. Set the needle and its float on top of the water.
6. Watch the needle move! It will line up with the north-south direction.

Robert Peary and a team of explorers carried a compass to the North Pole in 1909. You may not be headed to the North Pole, but you can still carry a compass made of household items. Then you'll be ready to do your own exploring!

Try this next!

Put the magnet from Step 1 near the bowl. Observe what the needle does. What does this tell you about the strength of the magnet's field compared with the strength of Earth's magnetic field?

QUESTION TIME!

What difficulties might explorers face if they tried to use a floating compass needle like this one to find their way on a trek or a sailing ship?

45

Index

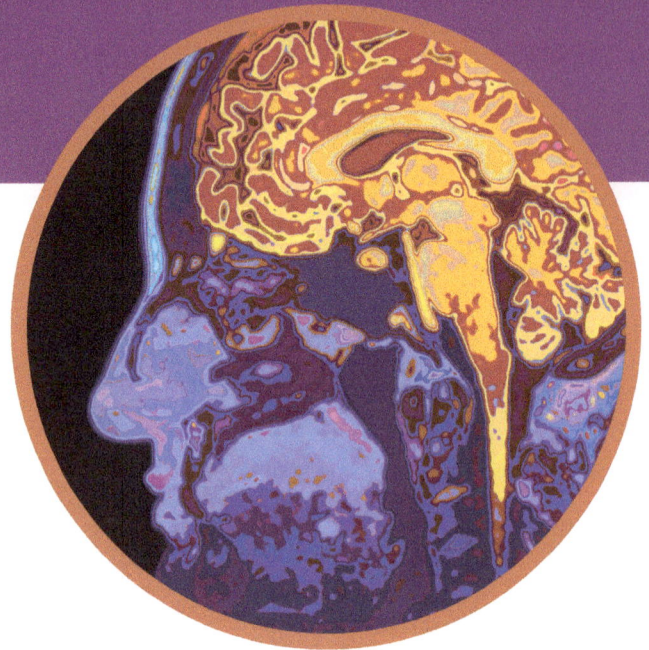

Glossary

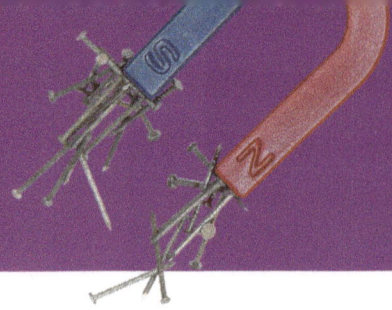

atom (AT um)—the smallest unit of matter

blackout (BLAK owt)—an event in which all lights are off due to an electrical power failure

electromagnetism (ee LEK tro MAG nuh tih zem)—the coexistence of electric and magnetic fields

electron (ee LEK trahn)—a negatively charged particle that moves around the core of an atom

exoplanet (EK so plae net)—a planet outside our solar system

force (FORS)—a push or a pull exerted on an object

fusion (FYU shun)—the uniting of two atoms under extreme pressure and temperature, resulting in the release of enormous amounts of energy

gravity (GRAE vih tee)—the force that draws objects toward the center of a planet or other large body

levitation (leh vih TAY shun)—the act of rising into or floating in the air

magnetic domain (mag NEH tik doe MAYN)—a region in which the magnetic fields of atoms are grouped together and aligned

magnetoreception (mag NEH toh ree SEP shun)—the ability to sense magnetic fields

mass (MAS)—a measure of the amount of matter in an object

Mesoamerica (meh zoe uh MARE ih kah)—a region including parts of modern-day Mexico and Central America that was historically occupied by various Indigenous (native) cultures

radiation (ray dee AY shun)—energy that spreads out from a source in the form of waves

resistance (ree ZIH stents)—a measurement of how difficult it is for an electric current to pass through a circuit

volt (VOHLT)—a unit of measurement for electric current

www.ingramcontent.com/pod-product-compliance
Lightning Source LLC
Chambersburg PA
CBHW040144200326
41519CB00032B/7594